PROTONS NEUTRONS ELECTRONS
Physics Kids
Children's Physics Books Education

SPEEDY
PUBLISHING

Speedy Publishing LLC
40 E. Main St. #1156
Newark, DE 19711
www.speedypublishing.com

The discoveries of electrons, protons and neutrons were made by J. J. Thomson, Ernest Rutherford and James Chadwick.

A proton
is part of
an atom.

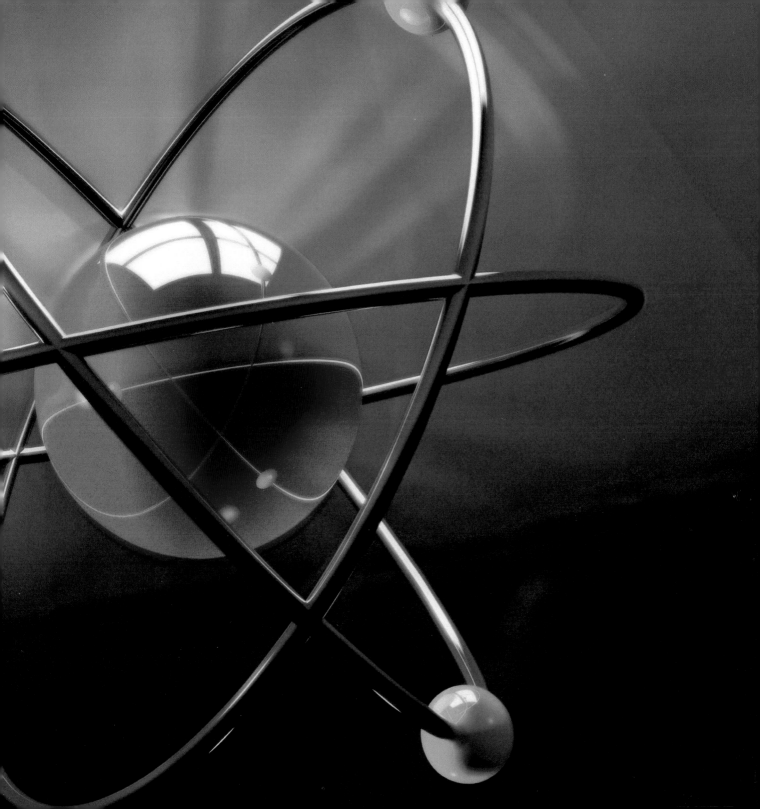

The proton is a positively charged particle that is located at the center of the atom in the nucleus.

All protons everywhere in the universe, are exactly the same.

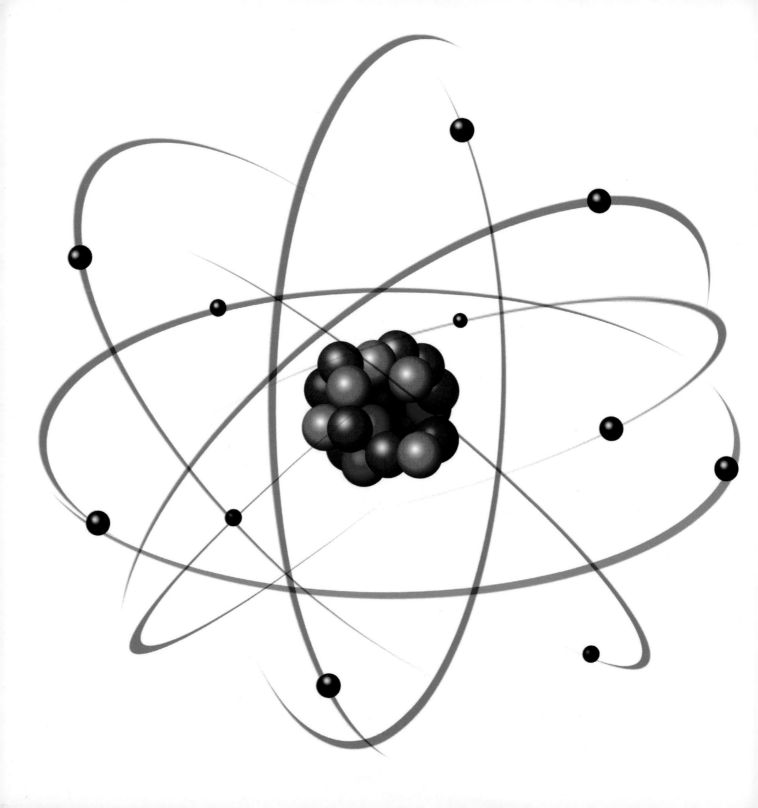

Different atoms
have different
numbers of protons.

Protons are actually made of even smaller invisible particles, called quarks.

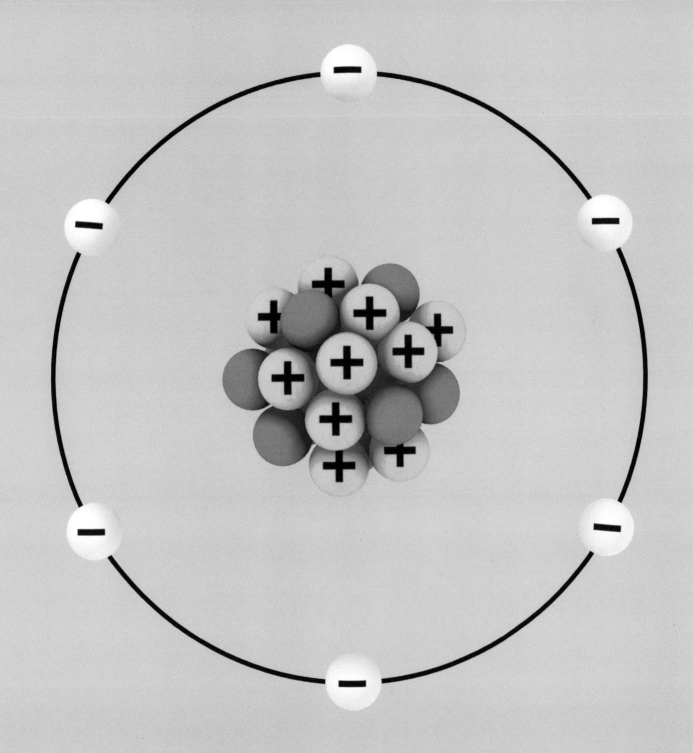

The mass of the proton is about 1,840 times the mass of the electron.

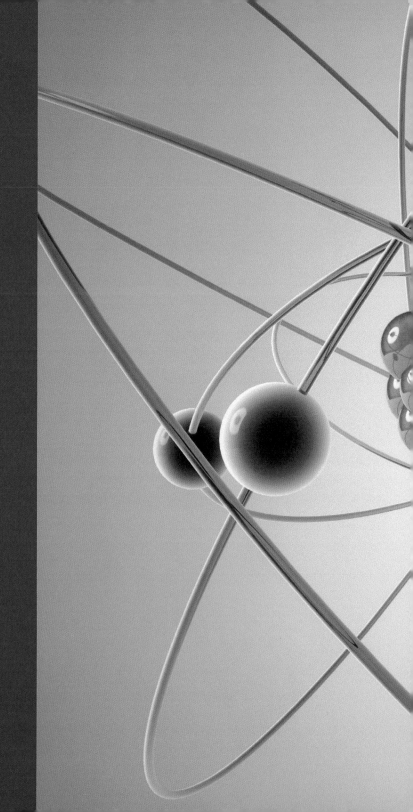

The neutron is a subatomic particle with a mass slightly larger than a proton.

Protons and neutrons constitute the nucleus of an atom.

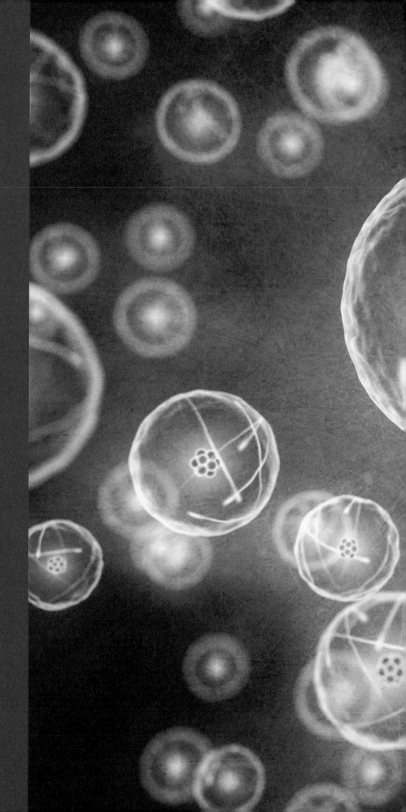

Unlike protons which have a positive charge, neutrons have zero charge.

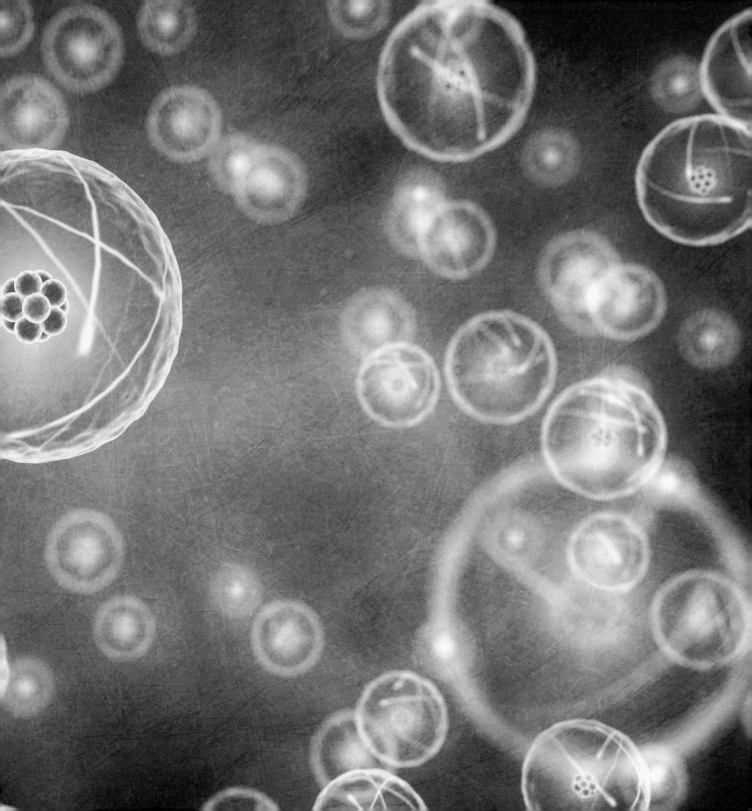

Neutrons are 1839 times heavier than electrons.

Neutrons are the key to nuclear chain reactions, nuclear power and nuclear weapons.

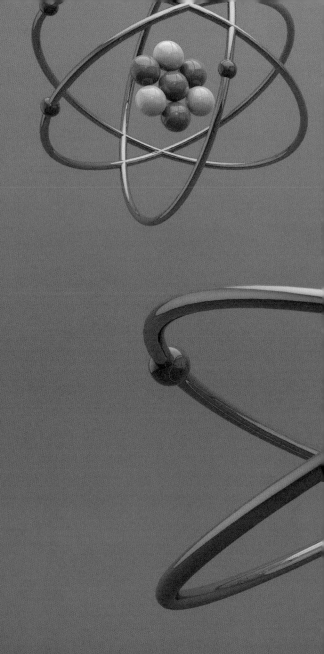

Neutrons are produced copiously in nuclear fission and fusion.

An electron is a very small piece of matter and energy.

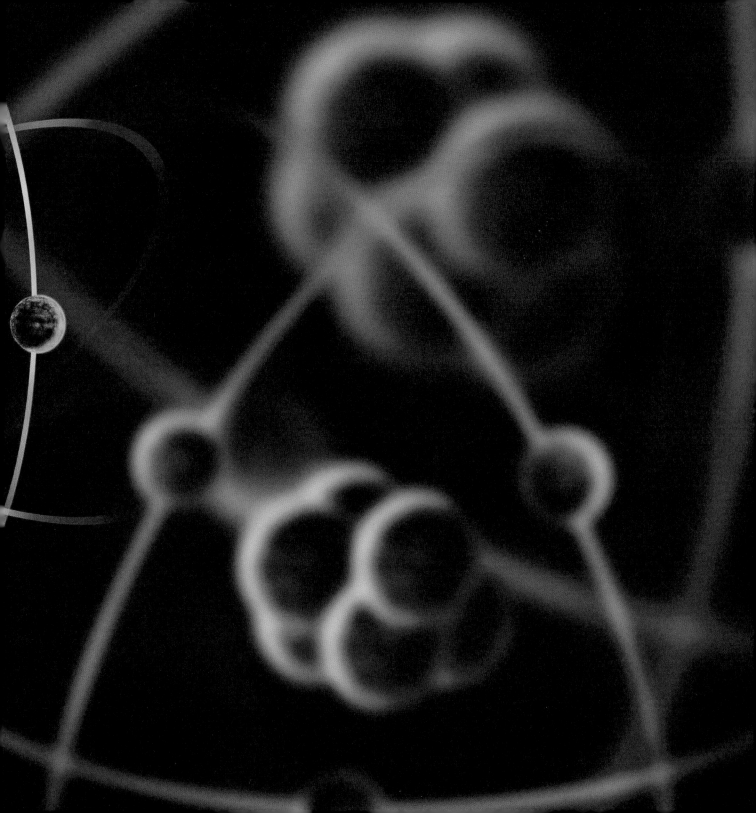

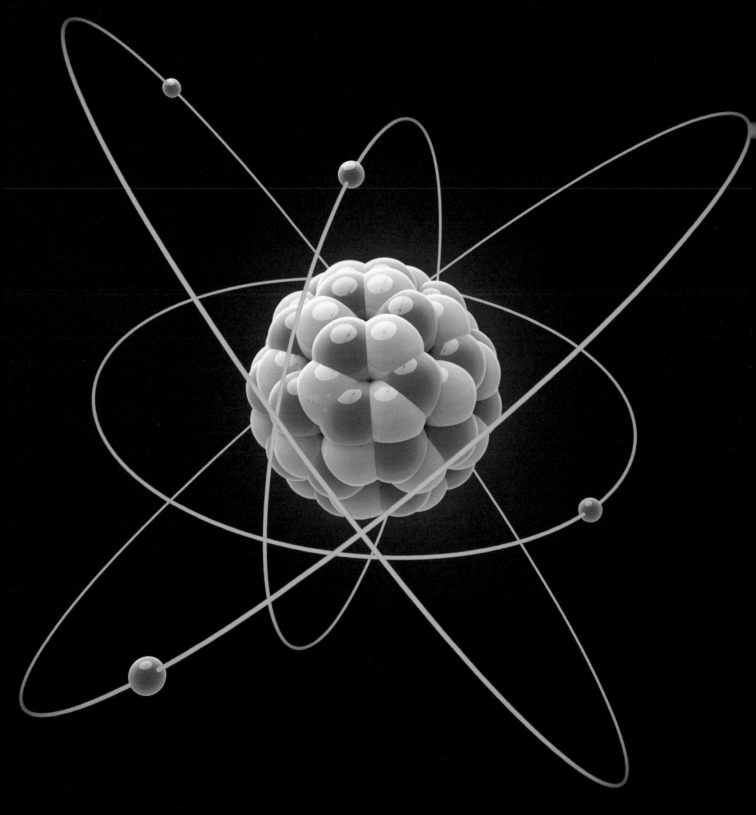

The electron is a subatomic particle with a negative electric charge.

Electrons have the smallest electrical charge.

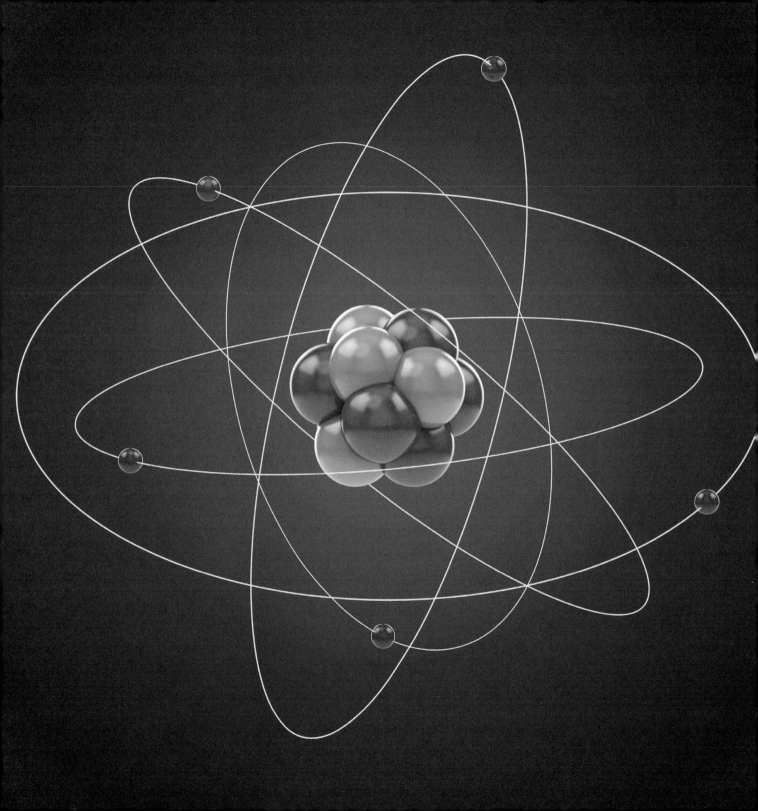

The electricity that powers radios, motors, and many other things consists of many electrons moving through wires.

Protons, Neutrons and Electrons form an atom.

The atom
is the basic
building
block for all
matter in
the universe.

Printed in Great Britain
by Amazon